QUARTER MILLION, QUARTER MILE
THE REAL STORY OF THE LEGENDARY STREET RACE

ROBERT BLACKWELL

Copyright © 2020 by Robert Blackwell

All rights reserved.

No part of this book may be reproduced in any form or by any electronic or mechanical means, including information storage and retrieval systems, without written permission from the author, except for the use of brief quotations in a book review.

For my Drag Racing Family

INTRODUCTION

On television today there are many shows about street racing. There are The Fastest Cars in the Dirty South. The Discovery Channel has it's own Street Outlaws franchise with the Street Outlaws, the group from Oklahoma, Street Outlaws: Memphis, and there are probably more shows that I don't know about. The Oklahoma show was at one time the number one show on the Discovery Channel. There are street racing teams from Mississippi, Kentucky, Detroit, Chicago, South Carolina, and more. These shows have a colorful cast of characters like Big Chief, Eric "the Prosecutor", Daddy Dave, Barefoot Ronnie and JJ da Boss.

But big-time street racing was born in New York in the 1960's on the Connecting Highway. There were characters like Rapid Ronald Lyles, John "Mutt" Lyles, Eugene "Bonanza" Coard, Tab Talmadge, "Super John" McFadden, "Fast Earl" Mitchell, Willie "Cam Rod" Campbell, The Smallwood Brothers, Puerto Rican John Sandoval, Levi "The Black Knight" Holmes, and Rufus "Brooklyn Heavy" Boyd

INTRODUCTION

The Gathering Places

There were many places racers and spectators would gather to talk about their cars and set up races back then.

The White Castle Hamburger joint on Atlantic Avenue in Brooklyn was a favorite hangout spot. There always seemed to be a crowd at J&B Automotive at 895 Bedford Avenue, a gas station owned by Jesse Johnson and Bennie Dunham and something was always taking place in the parking lot of Mitchell's hamburger restaurant on 7th Avenue in Brooklyn, where people would gather to set-up their runs and then they would go under the Brooklyn-Queens Expressway, or to Second Avenue to race. In Queens they raced Cross Bay Boulevard, Connecting Highway and Nassau Expressway. Cross Bay Boulevard was an empty strip of highway with nothing but swamp stretching for miles on each side of the highway.

One of the most amazing things about the illegal street racing in New York was that someone would print flyers to get the word out about the big street races scheduled to occur.

The New York Street Racing Scene

The New York Street Racing Scene

Big time street racing was born in New York on the Connecting Highway. This street connects the Brooklyn-Queens Expressway to the Grand Central Parkway in Queens, New York. This was such a great area because from one underpass to the next underpass was a quarter-mile. This was one of the favorite spots for big money races. Most of the races took place after midnight because the highway was usually deserted by local traffic and was less dangerous. The

INTRODUCTION

racing took place in summer and winter, any season because they were racers.

I don't know what year street racing actually began in New York but it was very popular in the 1960's and 70's. Street racing became so big in the 60's and 70's that venders began to set up on the streets selling hotdogs, drinks and ice cream to the crowds.

The Lyles Brothers and crew chief Eugene Coard and Bennie Dunham, Jessie Johnson, and a young eight-year-old Allen Coard, Eugene's son made the nightly rounds of the New York street-racing circuit. They raced the Clearview Expressway, the Connecting Highway, 150th St., near the Kennedy Airport Service Road, Fountain Ave, the service road off the Long Island Expressway, and just about any stoplight with a couple of blocks of clear pathway ahead.

The Mutt Brothers team raced a 1968 Plymouth GTX with a 426 Hemi engine on the street. While testing the GTX at the West Hampton drag strip on Long Island, Brian Higgins the owner of S&K Speed Shop Long Island New York, was there driving a 1968 Dodge Dart factory Super Stock car. Chrysler Corporation only made 80 of them specifically for the NHRA Super Stock/B class. He was cruising down the track going through the 4-speed gearbox and running faster than the Mutt Brothers Plymouth GTX by almost a second. Eugene walked over to him and asked him if the car for sale, he said yes for $6,000. And everything after that is history. Each team member chipped in $1,200, and the next week, they took the money over to S&K Speed Shop and became the proud owners of the 1968 Dodge Dart factory Super Stock car.

S&K Speed Shop painted the car black, and before the paint was dry, the team had taken the car to the streets of New York and made

INTRODUCTION

their money back and more. Because the Dart was not known on the streets, the rumor was that the Mutt Brothers had purchased a stock Dart took the motor out the 68 Hemi GTX and put it in the Dodge Dart with 10-inch tires and no traction bars and was looking for opponents to race. The same night they raced the car four times and won each time, beating a 327 Corvette, a 396 Chevelle, a Plymouth GTX with a 426 Hemi, a 396cubic inch-375 horsepower Camaro. The Mutt Brothers raced the 1968, Dodge Dart, two to three times a week. The Dart was undefeated on the streets and never went through the quarter, in a race, under full power. That was how strong that factory 1968 Dodge Dart was on the street. After winning so many races and going undefeated, there were rumors circulating that the Dart was being tuned by Jake King from the Sox and shop in Burlington, North Carolina. However, that was not true. The car was never taken to North Carolina, and Jake King never came to New York to tune the Dart. The car was run the way it was purchased from Brian Higgins at S&K Speed Shop other than valve and clutch adjustments.

The John McFadden Call

Later as the fame of the Dodge Dart spread, everyone wanted to beat that Black Dart. So the competition got stronger, and the bets got higher. Super John McFadden called the crew chief Eugene Coard and wanted to race. John had a 69 Camaro 396 cubic inch-375 horsepower that Mutt had beaten on the streets before. Of all the races the Dart had on the street, McFadden had come closest to beating the Dart but not knowing the Dart had never run through the quarter, in a race, under full power. McFadden felt that if he upgraded his car, he could win the race. So he went to the man known as "Mr. Chevrolet," Dick Harrell who rebuilt that car and installed 488 cubic inch super light big block Chevrolet engine and

INTRODUCTION

lightened the car up to match race weight and that's when Super John felt he was ready for the big race. The legendary race was called the quarter million quarter mile race.

The Connecting Highway

Everyone knew that all the big money races took place on the Connecting Highway. All the big-name street racers knew this, and so did the NYPD, New York's finest. The main reason most real street racers liked the Connecting Highway was that from one underpass to the next underpass was a quarter of a mile. The spectators would all pack in to see the races because you could see everything, so if you wanted to see a good race, the Connecting Highway was the place.

At the Connecting Highway all the racers used the two elevated service roads on each side of the highway as the pits to work on their cars. People would do all types of repairs and services to cars. It kind of reminds you of a modern-day NASCAR pit when you saw how fast the work was done, but remember this was a public street and in the middle of the night. Street racing was then and even today an adrenaline rush, and there was big money to be made. So to see the quick work being done on the race cars was an even greater adrenaline rush.

To watch the races, the crowd looked down onto the highway from the two guardrails that ran along the elevated service roads.

The New York Police Department would hide and watch cars go by with open headers and racing slicks, making their way to the race site. When the police became tired of handling the situation or the crowds became too large, they would call in the Fire Department of

INTRODUCTION

New York who would come to the Connecting Highway to open the fire hydrants and spray down the Connecting Highway from above. Sometimes this would happen every weekend and some weeknights.

WHO WERE THESE GUYS?

Who were these guys?

John "Mutt" Lyles, was from Bennettsville, South Carolina, an Army veteran, and excellent driver, a machinist at a local Brooklyn, New York Machine Shop, a motorcycle enthusiast, and the older brother of Rapid Ronald Lyles. Mutt and Ronald spent their days focused on automotive engines, including blocks and cylinder heads.

RAPID RONALD LYLES was from Bennettsville, South Carolina, worked as a machinist at a local Brooklyn Machine Shop, an excellent driver but not as good as his older brother. Later he becomes a regular in the National Hot Rod Association (NHRA), International Hot Rod Association (IHRA), and American Hot Rod Association (AHRA) Pro Stock class. He was the first and only African American member of the United States Racing Team, an elite group of the fastest Pro Stock racers in the United States.

. . .

Rufus "Brooklyn Heavy" Boyd was a street racer from Washington, North Carolina who lived in Brooklyn, NY. And ran a business called "J&L Racing Enterprises". "J&L Racing Enterprises" was located at 171 Lexington Avenue, in the Bedford-Stuyvesant section of Brooklyn. Heavy had many Chevrolet race cars and a few Mopars also.

Levi Holmes, lived in Newark, New Jersey and made a name for himself on the street driving a 1968 Chevrolet Camaro that was called the "Black Knight". Levi raced Pro Stock in his '68 Black Knight Camaro, then raced a black 1970 Nova SS with the "Black Knight" painted on the doors. Levi also owned the ex-MiMi 1969 Chevrolet Camaro. When not driving his personal cars Levi drove for "Brooklyn Heavy". Levi was featured in the January 1971 issue of Hi-Performance Cars magazine.

Tab Talmadge, was from Brooklyn, NY. Tab was a thriving street racer and lottery businessman. He purchased 'Dyno' Don Nicholson's 1965 Ford Mustang 427-ci SOHC 'Cammer' A/FX 4-speed car right after Don set a track record at Englishtown, New Jersey with the car. Tab was a real Ford man, and apparently one hell of a driver in any Ford he drove and reportedly some street racers would hire Tab to drive their cars for them. Tab passed away in December of 2007.

"Super John" McFadden was from Brooklyn, New York. McFadden, also known as "Big John", owned a 1969 Chevrolet Camaro that he had purchased from Dickie Harrell and later modified it into an SS/AA legal racer. He raced John "Mutt" Lyles of the Mutt Brothers in possibly the most famous street race to ever occur. This race was called the "Quarter Million Quarter Mile". He also raced the car in Pro Stock.

"Fast Earl" Mitchell

Fast Earl was from Paterson, NJ. 'Fast Earl' street raced on 150th in Brooklyn (near Kennedy Airport) and on Route 22 in Newark. He raced professionally in Pro Stock during the early-'70s. He owned a 1969 Chevrolet Camaro (which he used to race on the street and the track). Some sources state the car had a Booth-Arons built big-block and some even state Wally Booth actually owned and raced the car before Fast Earl Mitchell bought it. He was also a member of the "United Soul Racing Team", a team of racers created and organized by Mutt Brother member Eugene Coard.

Willie "Cam Rod" Campbell was originally from North Carolina but lived in Brooklyn, New York. Cam Rod raced in the streets of New York and later raced professionally in Pro Stock after he bought the Platt & Yates Pro Stock 1970 Ford Maverick from Pro Stock racer Hubert Platt from Georgia.

The Smallwood Brothers, James and Wilbert "Wicked Will" Smallwood raced both on the street and professionally in Pro Stock. They owned three ex-Sox & Martin race cars: a 1969 notchback Barracuda and a 1970 Plymouth Barracuda and a 1973 Plymouth Duster. The Smallwood Brothers street raced and match raced. In the summer of 1971, they traded the notchback Barracuda back to Sox & Martin for a 1970 Pro Stock Barracuda. But the Smallwood Brothers split with James racing a white 1st generation Camaro and Wicked Will" Smallwood purchasing the 1973 Plymouth Duster from Sox and Martin and continuing his pursuit of a championship in Pro Stock. I still remain in contact with Wilbert who no longer races.

PUERTO RICAN JOHN SANDOVAL was and still is a Bushwick, Brooklyn, New York street racer who was an all-around mechanical genius. John built engines, chassis, headers, a great driver, and the go-to man if you wanted your car to run fast.

BIG WILLIE ROBINSON the early 1970's legendary street racer, and the undisputed king of the late '60s- '70s of the East Los Angeles California street racing scene, and the founder and president of the International Brotherhood of Street Racers towed his 1969 Hemi Dodge Daytona Charger from Los Angeles, California to New York to race Ronald Lyles and the Mutt Brothers. The race never took place when the muscular 6'6" Vietnam veteran Big Willie and his wife Tomiko, realized that the Mutt Brothers car was a 1970 Sox & Martin Hemi Pro Stock Barracuda.

THE GATHERING PLACES

There were many places racers and spectators would gather to talk about their cars and set up races back then.

THE WHITE CASTLE Hamburger joint on Atlantic Avenue in Brooklyn was a favorite hangout spot. There always seemed to be a crowd at J&B Automotive at 895 Bedford Avenue, a gas station owned by Jesse Johnson and Bennie Dunham and something was always taking place in the parking lot of Mitchell's hamburger restaurant on 7th Avenue in Brooklyn, where people would gather to set-up their runs and then they would go under the Brooklyn-Queens Expressway, or to Second Avenue to race. In Queens they raced Cross Bay Boulevard, Connecting Highway and Nassau Expressway. Cross Bay Boulevard was an empty strip of highway with nothing but swamp stretching for miles on each side of the highway.

. . .

ROBERT BLACKWELL

ONE OF THE most amazing things about the illegal street racing in New York was that someone would print flyers to get the word out about the big street races scheduled to occur.

THE NEW YORK STREET RACING SCENE

Big time street racing was born in New York on the Connecting Highway. This street connects the Brooklyn-Queens Expressway to the Grand Central Parkway in Queens, New York. This was such a great area because from one underpass to the next underpass was a quarter-mile. This was one of the favorite spots for big money races. Most of the races took place after midnight because the highway was usually deserted by local traffic and was less dangerous. The racing took place in summer and winter, any season because they were racers.

I DON'T KNOW what year street racing actually began in New York but it was very popular in the 1960's and 70's. Street racing became so big in the 60's and 70's that venders began to set up on the streets selling hotdogs, drinks and ice cream to the crowds.

THE LYLES BROTHERS and crew chief Eugene Coard and Bennie Dunham, Jessie Johnson, and a young eight-year-old Allen Coard,

Eugene's son made the nightly rounds of the New York street-racing circuit. They raced the Clearview Expressway, the Connecting Highway, 150th St., near the Kennedy Airport Service Road, Fountain Ave, the service road off the Long Island Expressway, and just about any stoplight with a couple of blocks of clear pathway ahead.

THE MUTT BROTHERS team raced a 1968 Plymouth GTX with a 426 Hemi engine on the street. While testing the GTX at the West Hampton drag strip on Long Island, Brian Higgins the owner of S&K Speed Shop Long Island New York, was there driving a 1968 Dodge Dart factory Super Stock car. Chrysler Corporation only made 80 of them specifically for the NHRA Super Stock/B class. He was cruising down the track going through the 4-speed gearbox and running faster than the Mutt Brothers Plymouth GTX by almost a second. Eugene walked over to him and asked him if the car for sale, he said yes for $6,000. And everything after that is history. Each team member chipped in $1,200, and the next week, they took the money over to S&K Speed Shop and became the proud owners of the 1968 Dodge Dart factory Super Stock car.

S&K Speed Shop painted the car black, and before the paint was dry, the team had taken the car to the streets of New York and made their money back and more. Because the Dart was not known on the streets, the rumor was that the Mutt Brothers had purchased a stock Dart took the motor out the 68 Hemi GTX and put it in the Dodge Dart with 10-inch tires and no traction bars and was looking for opponents to race. The same night they raced the car four times and won each time, beating a 327 Corvette, a 396 Chevelle, a Plymouth GTX with a 426 Hemi, a 396cubic inch-375 horsepower Camaro. The Mutt Brothers raced the 1968, Dodge Dart, two to three times a week. The Dart was undefeated on the streets and never went through the quarter, in a race, under full power. That was how strong that factory 1968 Dodge Dart was on the street. After winning so many races and going undefeated, there were rumors circulating that the Dart was being tuned by Jake King from the Sox and shop in Burlington, North Carolina. However, that was not true. The car was never taken to North Carolina, and Jake King never came to New York to tune the Dart. The car was run the way it was purchased from Brian Higgins at S&K Speed Shop other than valve and clutch adjustments.

THE JOHN MCFADDEN CALL

*L*ater as the fame of the Dodge Dart spread, everyone wanted to beat that Black Dart. So the competition got stronger, and the bets got higher. Super John McFadden called the crew chief Eugene Coard and wanted to race. John had a 69 Camaro 396 cubic inch- 375 horsepower that Mutt had beaten on the streets before. Of all the races the Dart had on the street, McFadden had come closest to beating the Dart but not knowing the Dart had never run through the quarter, in a race, under full power. McFadden felt that if he upgraded his car, he could win the race. So he went to the man known as "Mr. Chevrolet," Dick Harrell who rebuilt that car and installed 488 cubic inch super light big block Chevrolet engine and lightened the car up to match race weight and that's when Super John felt he was ready for the big race. The legendary race was called the quarter million quarter mile race.

THE CONNECTING HIGHWAY

*E*veryone knew that all the big money races took place on the Connecting Highway. All the big-name street racers knew this, and so did the NYPD, New York's finest. The main reason most real street racers liked the Connecting Highway was that from one underpass to the next underpass was a quarter of a mile. The spectators would all pack in to see the races because you could see everything, so if you wanted to see a good race, the Connecting Highway was the place.

AT THE CONNECTING Highway all the racers used the two elevated service roads on each side of the highway as the pits to work on their cars. People would do all types of repairs and services to cars. It kind of reminds you of a modern-day NASCAR pit when you saw how fast the work was done, but remember this was a public street and in the middle of the night. Street racing was then and even today an adrenaline rush, and there was big money to be made. So to see the quick work being done on the race cars was an even greater adrenaline rush.

. . .

To watch the races, the crowd looked down onto the highway from the two guardrails that ran along the elevated service roads.

The New York Police Department would hide and watch cars go by with open headers and racing slicks, making their way to the race site. When the police became tired of handling the situation or the crowds became too large, they would call in the Fire Department of New York who would come to the Connecting Highway to open the fire hydrants and spray down the Connecting Highway from above. Sometimes this would happen every weekend and some weeknights.

THE RACE "QUARTER MILLION, QUARTER MILE"

There is a great legend about a quarter-mile, quarter-million-dollar race that took place between two teams. The two teams were the Mutt Brothers with the 1968 Dodge Dart Super Stock/B 426 Hemi 4-speed car that had just been repainted black vs. "Super John" McFadden in a white and red big block rat motored 427 with a tunnel ram intake 1969 Camaro A/Modified car.

I spoke with the Mutt Brothers crew chief Eugene Coard, and he stated that "people always ask me about this big money race, but it was nowhere near that amount of money people claim."

It was on a hot Tuesday night the first time The Quarter Million, Quarter Mile race was scheduled to take place. The cars lined up Mutt did a burnout, and the Dodge Dart dropped a driveshaft on the burn out so they couldn't race. On Wednesday, the second night, "Super John" McFadden's Camaro had trouble after the burnout with the transmission shifting so again they could not race. Each night the police officers were given one hundred dollars each, and they were satisfied. So on Tuesday night, the teams lost four hundred dollars, and on Wednesday night, they lost another four hundred dollars. The two groups were already eight hundred dollars in the hole and had not made one pass. By Thursday night, the news had traveled, and everyone had found out about the big race. People came from Staten Island, Long Island, Brooklyn, Queens, and other police officers. The teams told the policemen they only needed 10 seconds to go around and block off the areas so that "Super John" and Mutt could race. Mutt went up to the police officers and said, "give us a break tonight, we have already paid $800 if we run the race tonight we will pay you another $400 but if the race is not run then you just let us go and they agreed."

Here is how it actually happened that night. When the starter dropped his hand Super John in the Camaro got the hole shot and pulled a 1-foot high wheel stand for about 30 feet. For about an eighth of a mile the Camaro had a 2-car length lead over the Dodge Dart. Everyone thought the Dart would lose but once Mutt shifted into 3rd gear, the Black 426 Hemi Dart opened up and went flying by the Camaro winning the race by over two car lengths. The race was finally run, and the Mutt Brothers won.

After each race, the police officers would follow the cars through the quarter of a mile, and most people thought it was to arrest the drivers, but it was for them to get their money. Most of the police officers themselves were racers; however they raced at Westhampton Dragway on the weekends, and they knew that these street racers conducted the races in a safe manner. They made sure there were no cars on the street, and the races were always late at night.

THE RACERS WOULD ALWAYS SEEK to have a fair race. There was to be no jumping before the starter dropped his hands but concentrate on him and have a safe, clean race. Most racers did that, and there were no problems, but in case someone left early, and you followed the race was on, but if he leaves and you don't follow, then you won or he could come back and start all over.

THE RULES

Proverbs 4:5 says, "In all thy getting, get an understanding." Understanding was the greatest thing we could do when we were participating in a drag racing event, especially where are paying our money and that's why we had so many fair races because we always had a good understanding from the very beginning. All races were set up before we got to the race site, so when we arrived on site we knew exactly what each person was supposed to do. We would go there pull the two racecars to the line and from that point it was on, and as soon as the race was over we would go back to J&B Automotive or whatever site we setup the race.

ON THAT PARTICULAR THURSDAY NIGHT, there were people everywhere, and everyone considered that race to be the grand national of all street races, the world finals and at that time, that was the biggest race anyone had ever witnessed on Conduit Avenue. People were lined up all the way through the quarter of a mile, and probably another two-to-three people deep were at the starting line.

THE CONCLUSION

In the 1960's the Lyles brothers and crew chief Eugene Coard along with Benny and Jesse street raced every weekend and sometimes during the week. After Mutt was killed in a motorcycle accident, Eugene, Ronald and Bennie, left the streets regularly and set their sights on becoming professional racers. With the death of John "Mutt" Lyles in a motorcycle accident, they officially became the Mutt Brothers and opened several businesses together.

ONE OF THE trademarks of the Mutt Brothers was they bought the best equipment, their cars were always flawless in appearance and preparation, as was their tow vehicles, the team dressed in sharp in classy uniforms while other teams wore jeans and t-shirts, the Mutt Brothers went first class.

SOME OF THE CARS

1973 Pro Stock Duster

ROBERT BLACKWELL

United States Racing Team

AFTERWORD

The Mutt Brothers moved on to professional racing in the Pro Stock class in all three sanctioning bodies, NHRA, IHRA and AHRA.

ABOUT THE AUTHOR

Robert Blackwell is a retired labor relations consultant, pastor, speaker, and former high school teacher. Having a love for writing for many decades, he has written training manuals, articles for newsletters, journals, magazines and ten books. He lives in North Carolina and serves the local education and religious community. This is Robert's tenth book.

ALSO BY ROBERT BLACKWELL

As Long As I Live

Hole Shot: The Story of the Last Mutt Brother

TAPA (Timeless Wisdom and Practical Advice to Help Young Men NOW

The Book on African American Drag Racers

Study the Word: Bible Study and Prayer Journal

Dads Willing to be Dads

Mom I Remember: Exploring the Mom and Me Relationship

The Legendary Smallwood Brothers

A Love Supreme

The Gospel Singer

www.ingramcontent.com/pod-product-compliance
Lightning Source LLC
Chambersburg PA
CBHW050322220526
45465CB00005B/2094